Date:_____   T

Temperature:_____

Clouds: Cloudless   Partial   Covered

Precipitation: Fog   Mist   Rain   Hail
Thunderstorm   Frost   Snow

Wind: None Light Mild Strong Gusts

Ground: Dry Moist Wet Mud Flood
Frozen Ice Snow_____cm

Date:_____   Time:_____

Temperature:_____

[ ]

Clouds: Cloudless   Partial   Covered

_____

Precipitation: Fog   Mist   Rain   Hail
              Thunderstorm   Frost   Snow

_____

Wind: None Light Mild Strong Gusts

_____

Ground: Dry Moist Wet Mud Flood
        Frozen Ice Snow_____cm

Date:_____   Time:_____

Temperature:_____

[blank box]

Clouds: Cloudless   Partial   Covered

_____

Precipitation: Fog   Mist   Rain   Hail
        Thunderstorm   Frost   Snow

_____

Wind: None Light Mild Strong Gusts

_____

Ground: Dry Moist Wet Mud Flood
        Frozen   Ice   Snow_____cm

Date:_____   Time:_____

Temperature:_____

Clouds: Cloudless   Partial   Covered
_____

Precipitation: Fog   Mist   Rain   Hail
           Thunderstorm   Frost   Snow
_____

Wind: None Light Mild Strong Gusts
_____

Ground: Dry Moist Wet Mud Flood
         Frozen   Ice   Snow_____cm

Date:_____   Time:_____

Temperature:_____

Clouds: Cloudless Partial Covered

Precipitation: Fog Mist Rain Hail
Thunderstorm Frost Snow

Wind: None Light Mild Strong Gusts

Ground: Dry Moist Wet Mud Flood
Frozen Ice Snow_____cm

Date:_____   Time:_____

Temperature:_____

Clouds: Cloudless   Partial   Covered

_____

Precipitation: Fog   Mist   Rain   Hail
Thunderstorm   Frost   Snow

_____

Wind: None Light Mild Strong Gusts

_____

Ground: Dry Moist Wet Mud Flood
Frozen Ice Snow_____cm

Date:_____   Time:_____

Temperature:_____

Clouds: Cloudless   Partial   Covered

_____

Precipitation: Fog   Mist   Rain   Hail
              Thunderstorm   Frost   Snow

_____

Wind: None Light Mild Strong Gusts

_____

Ground: Dry Moist Wet Mud Flood
           Frozen   Ice   Snow_____cm

Date:_____   Time:_____

Temperature:_____

[ ]

Clouds: Cloudless   Partial   Covered
_____

Precipitation: Fog   Mist   Rain   Hail
Thunderstorm   Frost   Snow
_____

Wind: None Light Mild Strong Gusts
_____

Ground: Dry Moist Wet Mud Flood
Frozen   Ice   Snow_____cm

Date:_____   Time:_____

Temperature:_____

Clouds: Cloudless   Partial   Covered

Precipitation: Fog   Mist   Rain   Hail
Thunderstorm   Frost   Snow

Wind: None Light Mild Strong Gusts

Ground: Dry Moist Wet Mud Flood
Frozen Ice Snow_____cm

Date:_____   Time:_____

Temperature:_____

Clouds: Cloudless Partial Covered

_____

Precipitation: Fog Mist Rain Hail
Thunderstorm Frost Snow

_____

Wind: None Light Mild Strong Gusts

_____

Ground: Dry Moist Wet Mud Flood
Frozen Ice Snow_____cm

Date:_____   Time:_____

Temperature:_____

Clouds: Cloudless   Partial   Covered

_____

Precipitation: Fog   Mist   Rain   Hail
          Thunderstorm   Frost   Snow

_____

Wind: None Light Mild Strong Gusts

_____

Ground: Dry Moist Wet Mud Flood
          Frozen   Ice   Snow_____cm

Date:_____   Time:_____

Temperature:_____

Clouds: Cloudless   Partial   Covered

_____

Precipitation: Fog   Mist   Rain   Hail
            Thunderstorm   Frost   Snow

_____

Wind: None Light Mild Strong Gusts

_____

Ground: Dry Moist Wet Mud Flood
          Frozen   Ice   Snow_____cm

Date:_____   Time:_____

Temperature:_____

Clouds: Cloudless   Partial   Covered

_____

Precipitation: Fog   Mist   Rain   Hail
Thunderstorm   Frost   Snow

_____

Wind: None Light Mild Strong Gusts

_____

Ground: Dry Moist Wet Mud Flood
Frozen Ice Snow_____cm

Date:_____   Time:_____

Temperature:_____

Clouds: Cloudless   Partial   Covered

_____

Precipitation: Fog   Mist   Rain   Hail
Thunderstorm   Frost   Snow

_____

Wind: None Light Mild Strong Gusts

_____

Ground: Dry Moist Wet Mud Flood
Frozen Ice Snow_____cm

Date:_____   Time:_____

Temperature:_____

[ ]

Clouds: Cloudless   Partial   Covered

_____

Precipitation: Fog   Mist   Rain   Hail
Thunderstorm   Frost   Snow

_____

Wind: None Light Mild Strong Gusts

_____

Ground: Dry Moist Wet Mud Flood
Frozen   Ice   Snow_____cm

Date:_____   Time:_____

Temperature:_____

[ ]

Clouds: Cloudless   Partial   Covered

Precipitation: Fog   Mist   Rain   Hail
Thunderstorm   Frost   Snow

Wind: None   Light   Mild   Strong   Gusts

Ground: Dry   Moist   Wet   Mud   Flood
Frozen   Ice   Snow_____cm

Date:_____   Time:_____

Temperature:_____

Clouds: Cloudless   Partial   Covered

_____

Precipitation: Fog   Mist   Rain   Hail
           Thunderstorm   Frost   Snow

_____

Wind: None Light Mild Strong Gusts

_____

Ground: Dry Moist Wet Mud Flood
           Frozen   Ice   Snow_____cm

Date:_____   Time:_____

Temperature:_____

Clouds: Cloudless Partial Covered

_____

Precipitation: Fog Mist Rain Hail
Thunderstorm Frost Snow

_____

Wind: None Light Mild Strong Gusts

_____

Ground: Dry Moist Wet Mud Flood
Frozen Ice Snow_____cm

Date:_____ Time:_____

Temperature:_____

Clouds: Cloudless Partial Covered

_____

Precipitation: Fog Mist Rain Hail
Thunderstorm Frost Snow

_____

Wind: None Light Mild Strong Gusts

_____

Ground: Dry Moist Wet Mud Flood
Frozen Ice Snow_____cm

Date:_____   Time:_____

Temperature:_____

Clouds: Cloudless   Partial   Covered

_____

Precipitation: Fog   Mist   Rain   Hail
Thunderstorm   Frost   Snow

_____

Wind: None Light Mild Strong Gusts

_____

Ground: Dry Moist Wet Mud Flood
Frozen Ice Snow_____cm

Date:_____   Time:_____

Temperature:_____

Clouds: Cloudless Partial Covered

Precipitation: Fog Mist Rain Hail
Thunderstorm Frost Snow

Wind: None Light Mild Strong Gusts

Ground: Dry Moist Wet Mud Flood
Frozen Ice Snow_____cm

Date:_____   Time:_____

Temperature:_____

Clouds: Cloudless   Partial   Covered

_____

Precipitation: Fog   Mist   Rain   Hail
Thunderstorm   Frost   Snow

_____

Wind: None Light Mild Strong Gusts

_____

Ground: Dry Moist Wet Mud Flood
Frozen Ice Snow_____cm

Date:_____   Time:_____

Temperature:_____

Clouds: Cloudless   Partial   Covered

Precipitation: Fog   Mist   Rain   Hail
Thunderstorm   Frost   Snow

Wind: None Light Mild Strong Gusts

Ground: Dry Moist Wet Mud Flood
Frozen   Ice   Snow_____cm

Date:_____   Time:_____

Temperature:_____

Clouds: Cloudless Partial Covered

_____

Precipitation: Fog Mist Rain Hail
Thunderstorm Frost Snow

_____

Wind: None Light Mild Strong Gusts

_____

Ground: Dry Moist Wet Mud Flood
Frozen Ice Snow_____cm

Date:_____   Time:_____

Temperature:_____

Clouds: Cloudless   Partial   Covered

Precipitation: Fog   Mist   Rain   Hail
            Thunderstorm   Frost   Snow

Wind: None Light Mild Strong Gusts

Ground: Dry Moist Wet Mud Flood
            Frozen   Ice   Snow_____cm

Date:_____   Time:_____

Temperature:_____

Clouds: Cloudless   Partial   Covered

_____

Precipitation: Fog   Mist   Rain   Hail
Thunderstorm   Frost   Snow

_____

Wind: None Light Mild Strong Gusts

_____

Ground: Dry Moist Wet Mud Flood
Frozen   Ice   Snow_____cm

Date:_____   Time:_____

Temperature:_____

Clouds: Cloudless   Partial   Covered

Precipitation: Fog   Mist   Rain   Hail
        Thunderstorm   Frost   Snow

Wind: None Light Mild Strong Gusts

Ground: Dry Moist Wet Mud Flood
        Frozen Ice Snow_____cm

Date:_____   Time:_____

Temperature:_____

Clouds: Cloudless Partial Covered

_____

Precipitation: Fog Mist Rain Hail
Thunderstorm Frost Snow

_____

Wind: None Light Mild Strong Gusts

_____

Ground: Dry Moist Wet Mud Flood
Frozen Ice Snow_____cm

Date:_____   Time:_____

Temperature:_____

Clouds: Cloudless   Partial   Covered
_____

Precipitation: Fog   Mist   Rain   Hail
        Thunderstorm   Frost   Snow
_____

Wind: None Light Mild Strong Gusts
_____

Ground: Dry Moist Wet Mud Flood
        Frozen   Ice   Snow_____cm

Date:_____   Time:_____

Temperature:_____

Clouds: Cloudless   Partial   Covered

_____

Precipitation: Fog   Mist   Rain   Hail
        Thunderstorm   Frost   Snow

_____

Wind: None Light Mild Strong Gusts

_____

Ground: Dry Moist Wet Mud Flood
        Frozen Ice Snow_____cm

Date:_____ Time:_____

Temperature:_____

Clouds: Cloudless Partial Covered

_____

Precipitation: Fog Mist Rain Hail
Thunderstorm Frost Snow

_____

Wind: None Light Mild Strong Gusts

_____

Ground: Dry Moist Wet Mud Flood
Frozen Ice Snow_____cm

Date:_____   Time:_____

Temperature:_____

[blank boxed area]

Clouds: Cloudless Partial Covered
_____

Precipitation: Fog Mist Rain Hail
Thunderstorm Frost Snow
_____

Wind: None Light Mild Strong Gusts
_____

Ground: Dry Moist Wet Mud Flood
Frozen Ice Snow_____cm

Date:_____   Time:_____

Temperature:_____

Clouds: Cloudless   Partial   Covered

_____

Precipitation: Fog   Mist   Rain   Hail
Thunderstorm   Frost   Snow

_____

Wind: None   Light   Mild   Strong   Gusts

_____

Ground: Dry   Moist   Wet   Mud   Flood
Frozen   Ice   Snow_____cm

Date:_____   Time:_____

Temperature:_____

Clouds: Cloudless   Partial   Covered

_____

Precipitation: Fog   Mist   Rain   Hail
Thunderstorm   Frost   Snow

_____

Wind: None Light Mild Strong Gusts

_____

Ground: Dry Moist Wet Mud Flood
Frozen   Ice   Snow_____cm

Date:_____   Time:_____

Temperature:_____

Clouds: Cloudless   Partial   Covered

_____

Precipitation: Fog   Mist   Rain   Hail
          Thunderstorm   Frost   Snow

_____

Wind: None Light Mild Strong Gusts

_____

Ground: Dry Moist Wet Mud Flood
          Frozen   Ice   Snow_____cm

Date:_____ Time:_____

Temperature:_____

Clouds: Cloudless Partial Covered

_____

Precipitation: Fog Mist Rain Hail
Thunderstorm Frost Snow

_____

Wind: None Light Mild Strong Gusts

_____

Ground: Dry Moist Wet Mud Flood
Frozen Ice Snow_____cm

Date:_____   Time:_____

Temperature:_____

Clouds: Cloudless   Partial   Covered

_____

Precipitation: Fog   Mist   Rain   Hail
Thunderstorm   Frost   Snow

_____

Wind: None Light Mild Strong Gusts

_____

Ground: Dry Moist Wet Mud Flood
Frozen   Ice   Snow_____cm

Date:_____   Time:_____

Temperature:_____

Clouds: Cloudless   Partial   Covered

_____

Precipitation: Fog   Mist   Rain   Hail
Thunderstorm   Frost   Snow

_____

Wind: None Light Mild Strong Gusts

_____

Ground: Dry Moist Wet Mud Flood
Frozen Ice Snow_____cm

Date:_____   Time:_____

Temperature:_____

Clouds: Cloudless Partial Covered

_____

Precipitation: Fog Mist Rain Hail
        Thunderstorm Frost Snow

_____

Wind: None Light Mild Strong Gusts

_____

Ground: Dry Moist Wet Mud Flood
        Frozen Ice Snow_____cm

Date:_____   Time:_____

Temperature:_____

Clouds: Cloudless   Partial   Covered
_____

Precipitation: Fog   Mist   Rain   Hail
          Thunderstorm   Frost   Snow
_____

Wind: None Light Mild Strong Gusts
_____

Ground: Dry Moist Wet Mud Flood
          Frozen   Ice   Snow_____cm

Date:_____   Time:_____

Temperature:_____

Clouds: Cloudless   Partial   Covered

_____

Precipitation: Fog   Mist   Rain   Hail
          Thunderstorm   Frost   Snow

_____

Wind: None Light Mild Strong Gusts

_____

Ground: Dry Moist Wet Mud Flood
          Frozen   Ice   Snow_____cm

Date:_____   Time:_____

Temperature:_____

Clouds: Cloudless   Partial   Covered
_____

Precipitation: Fog   Mist   Rain   Hail
Thunderstorm   Frost   Snow
_____

Wind: None Light Mild Strong Gusts
_____

Ground: Dry Moist Wet Mud Flood
Frozen   Ice   Snow_____cm

Date:_____   Time:_____

Temperature:_____

Clouds: Cloudless Partial Covered

Precipitation: Fog Mist Rain Hail
Thunderstorm Frost Snow

Wind: None Light Mild Strong Gusts

Ground: Dry Moist Wet Mud Flood
Frozen Ice Snow_____cm

Date:_____   Time:_____

Temperature:_____

Clouds: Cloudless   Partial   Covered
_____

Precipitation: Fog   Mist   Rain   Hail
            Thunderstorm   Frost   Snow
_____

Wind: None Light Mild Strong Gusts
_____

Ground: Dry Moist Wet Mud Flood
            Frozen Ice Snow_____cm

Date:_____   Time:_____

Temperature:_____

Clouds: Cloudless   Partial   Covered

_____

Precipitation: Fog   Mist   Rain   Hail
          Thunderstorm   Frost   Snow

_____

Wind: None Light Mild Strong Gusts

_____

Ground: Dry Moist Wet Mud Flood
          Frozen   Ice   Snow_____cm

Date:_____   Time:_____

Temperature:_____

Clouds: Cloudless   Partial   Covered

_____

Precipitation: Fog   Mist   Rain   Hail
Thunderstorm   Frost   Snow

_____

Wind: None   Light   Mild   Strong   Gusts

_____

Ground: Dry   Moist   Wet   Mud   Flood
Frozen   Ice   Snow_____cm

Date:_____   Time:_____

Temperature:_____

Clouds: Cloudless   Partial   Covered

_____

Precipitation: Fog   Mist   Rain   Hail
Thunderstorm   Frost   Snow

_____

Wind: None Light Mild Strong Gusts

_____

Ground: Dry Moist Wet Mud Flood
Frozen   Ice   Snow_____cm

Date:_____   Time:_____

Temperature:_____

Clouds: Cloudless Partial Covered

Precipitation: Fog Mist Rain Hail
Thunderstorm Frost Snow

Wind: None Light Mild Strong Gusts

Ground: Dry Moist Wet Mud Flood
Frozen Ice Snow_____cm

Date:_____ Time:_____

Temperature:_____

Clouds: Cloudless Partial Covered

Precipitation: Fog Mist Rain Hail
Thunderstorm Frost Snow

Wind: None Light Mild Strong Gusts

Ground: Dry Moist Wet Mud Flood
Frozen Ice Snow_____cm

Date:_____   Time:_____

Temperature:_____

Clouds: Cloudless   Partial   Covered

_____

Precipitation: Fog   Mist   Rain   Hail
            Thunderstorm   Frost   Snow

_____

Wind: None Light Mild Strong Gusts

_____

Ground: Dry Moist Wet Mud Flood
            Frozen Ice Snow_____cm

Date:_____   Time:_____

Temperature:_____

Clouds: Cloudless Partial Covered

Precipitation: Fog Mist Rain Hail
        Thunderstorm Frost Snow

Wind: None Light Mild Strong Gusts

Ground: Dry Moist Wet Mud Flood
        Frozen Ice Snow_____cm

Date:_____   Time:_____

Temperature:_____

Clouds: Cloudless   Partial   Covered
_____

Precipitation: Fog   Mist   Rain   Hail
          Thunderstorm   Frost   Snow
_____

Wind: None Light Mild Strong Gusts
_____

Ground: Dry Moist Wet Mud Flood
          Frozen   Ice   Snow_____cm

Date:_____   Time:_____

Temperature:_____

Clouds: Cloudless   Partial   Covered

_____

Precipitation: Fog   Mist   Rain   Hail
        Thunderstorm   Frost   Snow

_____

Wind: None Light Mild Strong Gusts

_____

Ground: Dry Moist Wet Mud Flood
        Frozen   Ice   Snow_____cm

Date:_____   Time:_____

Temperature:_____

Clouds: Cloudless   Partial   Covered

_____

Precipitation: Fog   Mist   Rain   Hail
Thunderstorm   Frost   Snow

_____

Wind: None   Light   Mild   Strong   Gusts

_____

Ground: Dry   Moist   Wet   Mud   Flood
Frozen   Ice   Snow_____cm

Date:_____   Time:_____

Temperature:_____

Clouds: Cloudless   Partial   Covered

_____

Precipitation: Fog   Mist   Rain   Hail
         Thunderstorm   Frost   Snow

_____

Wind: None Light Mild Strong Gusts

_____

Ground: Dry Moist Wet Mud Flood
         Frozen   Ice   Snow_____cm

Date:_____   Time:_____

Temperature:_____

[ ]

Clouds: Cloudless   Partial   Covered
_____

Precipitation: Fog   Mist   Rain   Hail
          Thunderstorm   Frost   Snow
_____

Wind: None Light Mild Strong Gusts
_____

Ground: Dry Moist Wet Mud Flood
        Frozen   Ice   Snow_____cm

Date:_____ Time:_____

Temperature:_____

Clouds: Cloudless Partial Covered
_____

Precipitation: Fog Mist Rain Hail
    Thunderstorm Frost Snow
_____

Wind: None Light Mild Strong Gusts
_____

Ground: Dry Moist Wet Mud Flood
    Frozen Ice Snow_____cm

Date:_____ Time:_____

Temperature:_____

Clouds: Cloudless Partial Covered

_____

Precipitation: Fog Mist Rain Hail
Thunderstorm Frost Snow

_____

Wind: None Light Mild Strong Gusts

_____

Ground: Dry Moist Wet Mud Flood
Frozen Ice Snow_____cm

Date:_____   Time:_____

Temperature:_____

Clouds: Cloudless   Partial   Covered

_____

Precipitation: Fog   Mist   Rain   Hail
              Thunderstorm   Frost   Snow

_____

Wind: None Light Mild Strong Gusts

_____

Ground: Dry Moist Wet Mud Flood
        Frozen Ice Snow_____cm

Date:_____   Time:_____

Temperature:_____

[ ]

Clouds: Cloudless   Partial   Covered

_____

Precipitation: Fog   Mist   Rain   Hail
Thunderstorm   Frost   Snow

_____

Wind: None Light Mild Strong Gusts

_____

Ground: Dry Moist Wet Mud Flood
Frozen Ice Snow_____cm

Date:_____   Time:_____

Temperature:_____

Clouds: Cloudless   Partial   Covered

_____

Precipitation: Fog   Mist   Rain   Hail
         Thunderstorm   Frost   Snow

_____

Wind: None Light Mild Strong Gusts

_____

Ground: Dry Moist Wet Mud Flood
         Frozen   Ice   Snow_____cm

Date:_____   Time:_____

Temperature:_____

[ ]

Clouds: Cloudless   Partial   Covered
_____

Precipitation: Fog   Mist   Rain   Hail
          Thunderstorm   Frost   Snow
_____

Wind: None Light Mild Strong Gusts

_____

Ground: Dry Moist Wet Mud Flood
          Frozen   Ice   Snow_____cm

Date:_____   Time:_____

Temperature:_____

Clouds: Cloudless Partial Covered

_____

Precipitation: Fog Mist Rain Hail
Thunderstorm Frost Snow

_____

Wind: None Light Mild Strong Gusts

_____

Ground: Dry Moist Wet Mud Flood
Frozen Ice Snow_____cm

Date:_____   Time:_____

Temperature:_____

Clouds: Cloudless   Partial   Covered
_____

Precipitation: Fog   Mist   Rain   Hail
            Thunderstorm   Frost   Snow
_____

Wind: None Light Mild Strong Gusts
_____

Ground: Dry Moist Wet Mud Flood
            Frozen Ice Snow_____cm

Date:_____   Time:_____

Temperature:_____

[ ]

Clouds: Cloudless   Partial   Covered

_____

Precipitation: Fog   Mist   Rain   Hail
        Thunderstorm   Frost   Snow

_____

Wind: None Light Mild Strong Gusts

_____

Ground: Dry Moist Wet Mud Flood
        Frozen   Ice   Snow_____cm

Date:_____   Time:_____

Temperature:_____

Clouds: Cloudless Partial Covered

Precipitation: Fog Mist Rain Hail
Thunderstorm Frost Snow

Wind: None Light Mild Strong Gusts

Ground: Dry Moist Wet Mud Flood
Frozen Ice Snow_____cm

Date:_____   Time:_____

Temperature:_____

Clouds: Cloudless   Partial   Covered

_____

Precipitation: Fog   Mist   Rain   Hail
           Thunderstorm   Frost   Snow

_____

Wind: None Light Mild Strong Gusts

_____

Ground: Dry Moist Wet Mud Flood
        Frozen Ice Snow_____cm

Date:_____   Time:_____

Temperature:_____

```






```

Clouds: Cloudless   Partial   Covered

_____

Precipitation: Fog   Mist   Rain   Hail
          Thunderstorm   Frost   Snow

_____

Wind: None Light Mild Strong Gusts

_____

Ground: Dry Moist Wet Mud Flood
          Frozen   Ice   Snow_____cm

Date:_____   Time:_____

Temperature:_____

Clouds: Cloudless   Partial   Covered

Precipitation: Fog   Mist   Rain   Hail
Thunderstorm   Frost   Snow

Wind: None Light Mild Strong Gusts

Ground: Dry Moist Wet Mud Flood
Frozen   Ice   Snow_____cm

Date:_____   Time:_____

Temperature:_____

Clouds: Cloudless   Partial   Covered

_____

Precipitation: Fog   Mist   Rain   Hail
          Thunderstorm   Frost   Snow

_____

Wind: None Light Mild Strong Gusts

_____

Ground: Dry Moist Wet Mud Flood
          Frozen Ice Snow_____cm

Date:_____   Time:_____

Temperature:_____

Clouds: Cloudless   Partial   Covered

_____

Precipitation: Fog   Mist   Rain   Hail
        Thunderstorm   Frost   Snow

_____

Wind: None Light Mild Strong Gusts

_____

Ground: Dry Moist Wet Mud Flood
        Frozen   Ice   Snow_____cm

Date:_____ Time:_____

Temperature:_____

Clouds: Cloudless Partial Covered

_____

Precipitation: Fog Mist Rain Hail
Thunderstorm Frost Snow

_____

Wind: None Light Mild Strong Gusts

_____

Ground: Dry Moist Wet Mud Flood
Frozen Ice Snow_____cm

Date:_____   Time:_____

Temperature:_____

Clouds: Cloudless   Partial   Covered

_____

Precipitation: Fog   Mist   Rain   Hail
          Thunderstorm   Frost   Snow

_____

Wind: None Light Mild Strong Gusts

_____

Ground: Dry Moist Wet Mud Flood
          Frozen   Ice   Snow_____cm

Date:_____   Time:_____

Temperature:_____

Clouds: Cloudless   Partial   Covered

_____

Precipitation: Fog   Mist   Rain   Hail
        Thunderstorm   Frost   Snow

_____

Wind: None Light Mild Strong Gusts

_____

Ground: Dry Moist Wet Mud Flood
        Frozen   Ice   Snow_____cm

Date:_____   Time:_____

Temperature:_____

Clouds: Cloudless   Partial   Covered

_____

Precipitation: Fog   Mist   Rain   Hail
           Thunderstorm   Frost   Snow

_____

Wind: None Light Mild Strong Gusts

_____

Ground: Dry Moist Wet Mud Flood
        Frozen   Ice   Snow_____cm

Date:_____   Time:_____

Temperature:_____

Clouds: Cloudless   Partial   Covered

_____

Precipitation: Fog   Mist   Rain   Hail
           Thunderstorm   Frost   Snow

_____

Wind: None Light Mild Strong Gusts

_____

Ground: Dry   Moist   Wet   Mud   Flood
        Frozen   Ice   Snow_____cm

Date:_____   Time:_____

Temperature:_____

┌─────────────────────────────────────┐
│   │
│   │
│   │
│   │
│   │
│   │
│   │
└─────────────────────────────────────┘

Clouds: Cloudless   Partial   Covered

_____

Precipitation: Fog   Mist   Rain   Hail
        Thunderstorm   Frost   Snow

_____

Wind: None Light Mild Strong Gusts

_____

Ground: Dry Moist Wet Mud Flood
        Frozen   Ice   Snow_____cm

Date:_____   Time:_____

Temperature:_____

Clouds: Cloudless   Partial   Covered
_____

Precipitation: Fog   Mist   Rain   Hail
          Thunderstorm   Frost   Snow
_____

Wind: None Light Mild Strong Gusts
_____

Ground: Dry Moist Wet Mud Flood
          Frozen   Ice   Snow_____cm

Date:_____   Time:_____

Temperature:_____

[ ]

Clouds: Cloudless   Partial   Covered

_____

Precipitation: Fog   Mist   Rain   Hail
Thunderstorm   Frost   Snow

_____

Wind: None Light Mild Strong Gusts

_____

Ground: Dry Moist Wet Mud Flood
Frozen   Ice   Snow_____cm

Date:_____   Time:_____

Temperature:_____

Clouds: Cloudless   Partial   Covered

_____

Precipitation: Fog   Mist   Rain   Hail
Thunderstorm   Frost   Snow

_____

Wind: None Light Mild Strong Gusts

_____

Ground: Dry Moist Wet Mud Flood
Frozen   Ice   Snow_____cm

Date:_____   Time:_____

Temperature:_____

Clouds: Cloudless Partial Covered

_____

Precipitation: Fog Mist Rain Hail
Thunderstorm Frost Snow

_____

Wind: None Light Mild Strong Gusts

_____

Ground: Dry Moist Wet Mud Flood
Frozen Ice Snow_____cm

Date:_____   Time:_____

Temperature:_____

Clouds: Cloudless   Partial   Covered
_____

Precipitation: Fog   Mist   Rain   Hail
          Thunderstorm   Frost   Snow
_____

Wind: None Light Mild Strong Gusts
_____

Ground: Dry Moist Wet Mud Flood
          Frozen   Ice   Snow_____cm

Date:_____   Time:_____

Temperature:_____

Clouds: Cloudless   Partial   Covered

_____

Precipitation: Fog   Mist   Rain   Hail
        Thunderstorm   Frost   Snow

_____

Wind: None Light Mild Strong Gusts

_____

Ground: Dry Moist Wet Mud Flood
        Frozen Ice Snow_____cm

Date:_____   Time:_____

Temperature:_____

[ ]

Clouds: Cloudless   Partial   Covered

_____

Precipitation: Fog   Mist   Rain   Hail
          Thunderstorm   Frost   Snow

_____

Wind: None Light Mild Strong Gusts

_____

Ground: Dry Moist Wet Mud Flood
          Frozen Ice Snow_____cm

Date:_____ Time:_____

Temperature:_____

Clouds: Cloudless Partial Covered

_____

Precipitation: Fog Mist Rain Hail
Thunderstorm Frost Snow

_____

Wind: None Light Mild Strong Gusts

_____

Ground: Dry Moist Wet Mud Flood
Frozen Ice Snow_____cm

Date:_____   Time:_____

Temperature:_____

Clouds: Cloudless   Partial   Covered

_____

Precipitation: Fog   Mist   Rain   Hail
            Thunderstorm   Frost   Snow

_____

Wind: None Light Mild Strong Gusts

_____

Ground: Dry Moist Wet Mud Flood
            Frozen Ice Snow_____cm

Date:_____   Time:_____

Temperature:_____

Clouds: Cloudless   Partial   Covered

_____

Precipitation: Fog   Mist   Rain   Hail
            Thunderstorm   Frost   Snow

_____

Wind: None Light Mild Strong Gusts

_____

Ground: Dry Moist Wet Mud Flood
            Frozen Ice Snow_____cm

Date:_____   Time:_____

Temperature:_____

Clouds: Cloudless   Partial   Covered

_____

Precipitation: Fog   Mist   Rain   Hail
Thunderstorm   Frost   Snow

_____

Wind: None Light Mild Strong Gusts

_____

Ground: Dry Moist Wet Mud Flood
Frozen   Ice   Snow_____cm

Date:_____   Time:_____

Temperature:_____

Clouds: Cloudless   Partial   Covered

_____

Precipitation: Fog   Mist   Rain   Hail
        Thunderstorm   Frost   Snow

_____

Wind: None Light Mild Strong Gusts

_____

Ground: Dry Moist Wet Mud Flood
        Frozen   Ice   Snow_____cm

Date:_____   Time:_____

Temperature:_____

Clouds: Cloudless   Partial   Covered

_____

Precipitation: Fog   Mist   Rain   Hail
          Thunderstorm   Frost   Snow

_____

Wind: None Light Mild Strong Gusts

_____

Ground: Dry Moist Wet Mud Flood
          Frozen   Ice   Snow_____cm

Date:_____   Time:_____

Temperature:_____

[ ]

Clouds: Cloudless   Partial   Covered

Precipitation: Fog   Mist   Rain   Hail
            Thunderstorm   Frost   Snow

Wind: None Light Mild Strong Gusts

Ground: Dry Moist Wet Mud Flood
            Frozen Ice Snow_____cm

Date:_____   Time:_____

Temperature:_____

Clouds: Cloudless   Partial   Covered

_____

Precipitation: Fog   Mist   Rain   Hail
             Thunderstorm   Frost   Snow

_____

Wind: None Light Mild Strong Gusts

_____

Ground: Dry Moist Wet Mud Flood
          Frozen   Ice   Snow_____cm

Date:_____   Time:_____

Temperature:_____

Clouds: Cloudless Partial Covered

_____

Precipitation: Fog Mist Rain Hail
Thunderstorm Frost Snow

_____

Wind: None Light Mild Strong Gusts

_____

Ground: Dry Moist Wet Mud Flood
Frozen Ice Snow_____cm

Date:_____ Time:_____

Temperature:_____

Clouds: Cloudless Partial Covered

_____

Precipitation: Fog Mist Rain Hail
Thunderstorm Frost Snow

_____

Wind: None Light Mild Strong Gusts

_____

Ground: Dry Moist Wet Mud Flood
Frozen Ice Snow_____cm

Date:_____   Time:_____

Temperature:_____

[ ]

Clouds: Cloudless   Partial   Covered

_____

Precipitation: Fog   Mist   Rain   Hail
            Thunderstorm   Frost   Snow

_____

Wind: None Light Mild Strong Gusts

_____

Ground: Dry Moist Wet Mud Flood
            Frozen Ice Snow_____cm

Date:_____   Time:_____

Temperature:_____

Clouds: Cloudless Partial Covered
_____

Precipitation: Fog   Mist   Rain   Hail
        Thunderstorm   Frost   Snow
_____

Wind: None Light Mild Strong Gusts
_____

Ground: Dry Moist Wet Mud Flood
        Frozen Ice Snow_____cm

Date:_____   Time:_____

Temperature:_____

Clouds: Cloudless Partial Covered

_____

Precipitation: Fog Mist Rain Hail
      Thunderstorm Frost Snow

_____

Wind: None Light Mild Strong Gusts

_____

Ground: Dry Moist Wet Mud Flood
      Frozen Ice Snow_____cm

Date:_____   Time:_____

Temperature:_____

Clouds: Cloudless   Partial   Covered
_____

Precipitation: Fog   Mist   Rain   Hail
            Thunderstorm   Frost   Snow
_____

Wind: None Light Mild Strong Gusts
_____

Ground: Dry Moist Wet Mud Flood
        Frozen Ice Snow_____cm